Carey's First Day

Printed in China. Interior pages printed on FSC certified paper.

Written in Collaboration with The School of San Isidro de Bolas and The Felix Ruben School Students

Editor: Anna Ludeke
Illustrations: William McCaffrey
Photos: Dr. Caroline S. Rogers; Cathi Campbell, WCS
Resources: Wildlife Conservation Society
Map: Molly McCormick

Library of Congress Control Number: 2011930230
ISBN 13: 978-0-9842146-7-9
ISBN 10: 0-984214-67-4

The authors: Fabretto Children's Foundation students in San Isidro de Bolas

About this Book

Hello Friends!

This book was written by the Nicaraguan children who are pictured to the left. The educational opportunity that these children have is the result of the work of the Fabretto Children's Foundation.

Our young authors first studied about endangered hawksbill sea turtles using materials provided by the Wildlife Conservation Society. Using these lessons, they wrote their own stories about sea turtles. We then combined our favorite parts of their stories into **Carey's First Day**, adding some of our own ideas when needed to help the pieces fit together.

At the end of this book you can explore the world of hawksbill sea turtles, discover the culture of Nicaragua, and learn more about the authors of this story. You can also read about two women who are working to make positive changes in the world. Patricia Aragonés is a former lawyer who gave up that career to spearhead Fabretto's operations in Europe. Cynthia Lagueux has been working on sea turtle conservation for thirty years after beginning her career in the Peace Corps in Honduras.

The team at Dot-to-Dot Children's Books hopes you enjoy **Carey's First Day**. After reading the book, we invite you to write your own story about sea turtles and share it with us on our website.

Far to the south of the United States—past Mexico, past Guatemala, even past Honduras—lies a green and beautiful land called Nicaragua. This is the story of a sea turtle born on Nicaragua's coastline.

One night, Carey was born on a beautiful beach on an island off the coast of Nicaragua.

He had not one or two, but 149 brothers and sisters. They were all born as the sun set into the sky—crack! crack! CRACK!

The turtles blinked the sand from their eyes and climbed out of their nest into the moonlight.

Carey and his brothers and sisters shined their bald turtle heads and brushed off their shells. They smiled, hearing the gentle waves of the sea nearby.

One by one, they crawled across the sand toward the water. The turtles knew that this sea was to be their big new home.

As Carey crawled across the sand, he suddenly heard the call of a bird. In front of him, two huge birds sat watching the turtles. Near the shore, other birds swooped and grabbed up little crabs in their sharp beaks.

Carey knew right away that these were very, very dangerous birds. He waited, watching the birds and thinking about what to do.

Carey had to reach the water somehow, but he was very scared. As quickly as he could, he raced down the beach along with his brothers and sisters.

But turtles don't move very fast. All at once, a bird swooped down, grabbed Carey by the back flipper and flew up toward the sky. Carey wiggled and jiggled with all his might, trying to escape the bird's beak.

Finally, he squirmed away and felt himself falling. Carey landed back on the beach with a thud, dizzy and confused. He moved first in one direction, and then another, but could not tell if he was going anywhere at all.

"Where am I?" he thought.

Lights shone on Carey's face from far, far, far away.

"That must be the light of stars on the sea!" thought Carey, excited. With new energy, he moved as quickly as he could toward the light.

He crawled for a long time. Now light shone from the left and the right, and Carey was not sure which way to go. The ground slowly became hard under his tired flippers.

Then Carey realized he had made a mistake. This was not the light of stars on the sea at all, but something very different. Humans had built houses on the beach, all covered in bright lights.

Sadly, Carey realized he might never find the sea.

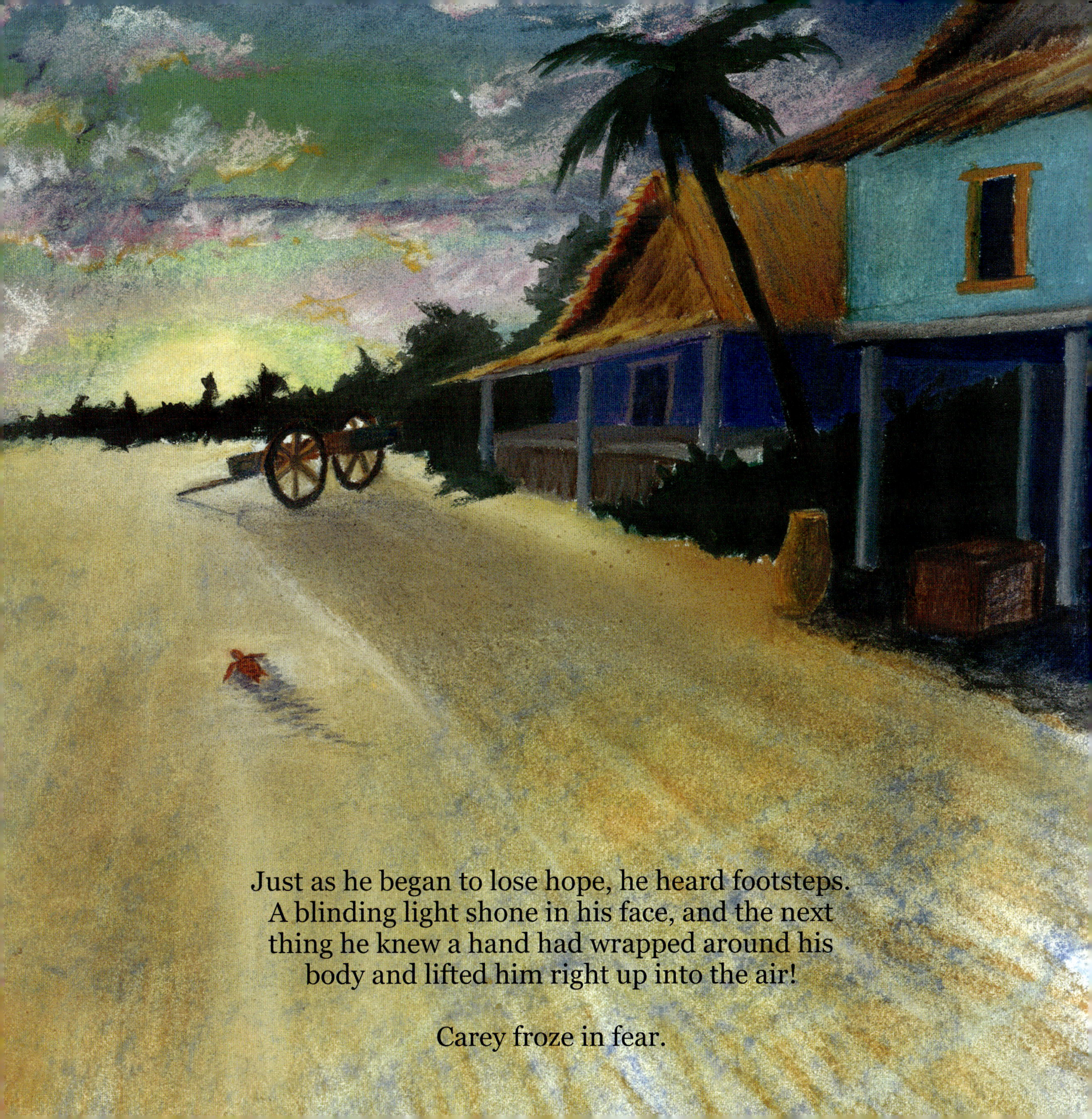

Just as he began to lose hope, he heard footsteps. A blinding light shone in his face, and the next thing he knew a hand had wrapped around his body and lifted him right up into the air!

Carey froze in fear.

His eyes were staring straight into the eyes of a human boy.

The boy said, "*Hola, tortuga*! You will make the perfect pet. I'm going to show you to mama and papa!"

With that, the boy dropped Carey into a bright orange bucket. He couldn't see a thing, and began to cry salty tears.

"I should never have tried to follow the lights," he cried. "If only I could find the sea and be home now."

The boy traveled by boat back to his town on the mainland, where his family earned a living selling goods at a local market.

During the trip, the boy gave Carey a new home in a large, wooden bucket.

Peering up out of his bucket, Carey saw that the boy had brought him to a market. People were just beginning to set up their stalls to sell their goods.

Carey waited and waited, afraid of what might happen. People passed noisily by his bucket, carrying many things in their arms.

Finally, Carey decided he had to escape. He didn't want to be eaten. He wanted to go home. But no matter how hard he pushed toward the top of the bucket, he kept slipping back to the bottom.

A girl came up to the stall and bent over the bucket.

"I think this turtle is trying to escape," she said. "How much for the little *tortuga*?"

The boy's face peered into the bucket. "He's my pet, but for twenty córdobas, you can have him!"

The girl pulled the money from her pocket and gave it to the boy.

Carey panicked, thinking to himself,
"Now this girl will eat me, and I will never return home."

But as the girl ran away from the market with Carey in her hand,
he heard a familiar sound: waves! Were they heading to the beach?
Could it be true?

The girl walked to the ocean and set Carey down next to the water.
"Now you are home and can live happily, little turtle," she said.
Carey was full of joy to be free again!

He felt so grateful to this kind girl who was patting him on the head. As she walked away, Carey read on the back of her shirt,

"D.E.T.E.—Don't Eat Turtle Eggs."

Now that Carey was in the sea, he was finally home. It had been an exciting first day of life but he was ready for a good rest.

As the sun began to set, Carey took a deep breath and let the rolling waves pull him into his big new home.

Hi Friends!

We hope you enjoyed this story. People from all over the world helped to bring it to life. But the story isn't over yet. You are now a part of it and we need your help.

When I am swimming together with my friends and family, we form a group called a bale. My friends at Dot-to-Dot Children's Books, Fabretto Children's Foundation, Facing the Future, and the Wildlife Conservation Society have joined my bale, and I hope that you will do the same.

You can join my bale by sharing the message of the Endangered Species, Empowered Communities (ESEC) stories with your friends and teachers. Whatever you do, taking action based on what you have learned about endangered species like me makes you a part of my team.

In the pages that follow this invitation, you can learn more about me and my friends. Just like the children who helped to write this book, you can use what you learn to write your own story about sea turtles and then share it with us at www.dot-to-dot-books.org.

I hope to hear from you soon!
Carey the Sea Turtle

Thank You!

My friends at Dot-to-Dot Children's Books and I want to express our sincere gratitude to the many people who helped us produce this book.

First and foremost, we want to recognize the creative contributions of the students who inspired and co-authored it. They are: María Fernanda Rivas, María Amalia Orozco, Kevin Reyes Zamora, Josseling Silva López, Rita Escorcia Gutiérrez, José Ariel Corea Zambrana, Jennifer Aracely Velásquez, Oriana Velásquez Corea, Leticia Silva López, Eliécer Mauricio Salgado Cabrera, Jennifer Silva López, Marlon Antonio Velásquez, Kennssy Isamar Picado Leiva, Jader Javier Jarquín, Josseling María Velásquez López, Jennifer María Alvarado Hernández, Ricardo José Bravo González, Henry Palacios, Ofilio Antonio Solís Jirón, Francisco Enrique Montiel Mercado, Freddy Ezequiel Leiva Bustamante, Israel Antonio Centeno Gutiérrez, Darwin Vallecillo Larios, Nancy del Socorro Maltez, Eveling Vasquez, Jean Carlos Maltez Martinez, Juan José Maltez, Mario Silva Lopez, Maria Isabel Perez, Vivian Ivonne Mora, Lene D'Trinidad, Abraham Murillo, Anielka del Rosario Gutierrez Rivas, Katherin Lugo, Fabricio Antonio Ordonez Munoz, José Maycol, Scarleth Paola, Jarvin, Ingris Palacio, Rosa Jenny Lanza, José Dolores, Helleng Martinez, Gabriel Alejandro Bustamante.

I also want to thank my other friends from around the world who helped make the story come to life: Patricia Aragones, Will Banham, Nic Buckley, Cathi Campbell, Christina Falck, Alexandra Garcia, Cynthia Lagueux, Anna Ludeke, Kevin Marinacci, Peter Schaller, Matthew Shea, and Jan Bachman of Devenish Press.

Many thanks to everyone! I appreciate everything that everyone has done for me!

Your Friend,
Carey the Sea Turtle

Hawksbill Sea Turtles

Hawksbill Sea Turtle Facts
Life span: 30-50 years in the wild
Weight & Size: 3 feet long and weighs about 150 pounds
Diet: omnivore - sea sponges, fish, mollusks, sea plants
Habitat: areas surrounding coral reefs

The hawksbill sea turtle has many adaptations that make it an excellent swimmer and hunter. Its wide flippers help it dive quickly down to coral reefs and then back up to the surface to get a breath of air. They also have a narrow beak that they use to probe cracks in coral reefs for their favorite food—sea sponges.

Female turtles go back to the exact beach where they were born to lay their eggs. In their nests, or "clutches," they can lay up to 150 eggs. They use their back flippers to dig the holes for their nests. After being buried, the eggs will take two months to hatch. Sea turtle babies are about as long as your thumb. When they are born, they walk straight to the open ocean and swim away. After a few years, they return to the coast to live in coral reefs.

Endangered: Hawksbill Sea Turtles

© Cathi Campbell, WCS

Turtle eggs are hunted by many predators, including dogs, cats, pigs, and rats. Crabs and birds also grab the babies as they make their dash over the sand to the sea. In the water, turtles have to watch out for hungry fish and diving birds that want to eat them. Sharks attack the turtle's flippers, hurting them so badly that they can not swim away.

Humans are the biggest threat to sea turtles. Sea turtles' beautiful shells are used for jewelry and their eggs are considered a delicacy to be eaten. Humans also destroy reefs where the sponges that turtles eat grow and they build on the beaches turtles use as nesting sites. Sea turtles sometimes drown in nets used by commercial fishermen.

© Dr. Caroline S. Rogers

By engaging local communities, schools, and governments, the Wildlife Conservation Society's Marine Turtle Conservation Program is helping to combat these threats to turtle populations in Nicaragua. Turn the page to learn more.

To learn more about the Wildlife Conservation Society and to support its efforts to protect wildlife and wild places worldwide, please visit www.wcs.org.

WILDLIFE CONSERVATION SOCIETY

Meet One of My Friends

Over the past fifteen years, the Wildlife Conservation Society (WCS) Sea Turtle Conservation Program in Nicaragua has made great strides in the conservation and recovery of sea turtle populations. More than 173,000 hawksbill turtle hatchlings have been produced through the program and Nicaraguan fishermen are now reporting sightings of small hawksbills in local waters.

Cynthia Lagueux has been working on sea turtle conservation and research issues for the past thirty years. She has consulted on sea turtle projects all over the world during a career that began in 1981 when she was a Peace Corps volunteer in Honduras. Working with colleagues, Cynthia developed and now directs the WCS Marine Turtle Conservation Program on the Caribbean coast of Nicaragua.

Inspired by her love of spending time in the natural world, Cynthia earned her doctoral degree in Wildlife Ecology and Conservation at the University of Florida. Her advice to children around the world who want to join conservation efforts is to realize that "the issues of conservation are complex and multifaceted" and require an understanding of culture as well as science.

Where is Nicaragua?

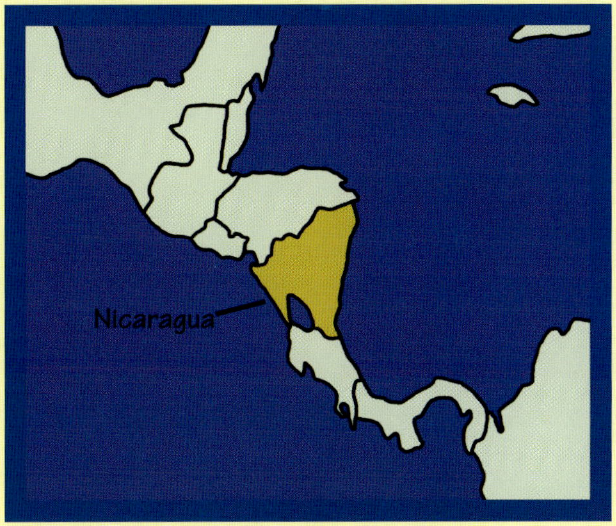

Nicaragua Facts
Continent: Central America
Capital City: Managua
Official Language: Spanish
Neighboring Countries: Costa Rica, Honduras

• Nicaragua is a country full of amazing wildlife, with animals like sloths, monkeys, and sea turtles. Nicaragua is one of the most important breeding places for turtles.

• One of the most popular meals in Nicaragua is rice and beans. In Spanish it's called gallo pinto.

• Coffee is Nicaragua's biggest export product. Most of the coffee is grown in the lush, mountainous northern central region of the country.

• Lake Nicaragua is one of the largest in Latin America. Some of the fish in the lake are usually found only in saltwater but they have adapted to the freshwater.

Life in Nicaragua

1. Most Nicaraguans are Christian and cathedrals and churches are found in cities and villages across the country. Some of the churches are more than 400 years old.
2. Local markets are common in Nicaragua and feature fresh produce and local crafts.
3. Nicaragua's roads are full of cars, trucks, and motorcycles, but people also use carts pulled by horses or cattle to transport goods to market.

Meet the Authors

Forty Nicaraguan third and fourth graders helped write **Carey's First Day**. Half of the students attend an elementary school named after their small community, San Isidro de Bolas. The other students attend a school named after the famous Nicaraguan poet and diplomat, Felix Ruben (Rubén Darío).

These forty students are among the 600 served by Fabretto in the area surrounding San Isidro. Fabretto Children's Foundation works with more than 10,000 children to help them try to end the cycle of poverty in their lives. Fabretto programs work well because they begin working with students in preschool and continue into young adulthood.

During the creative writing workshop that resulted in the writing of **Carey's First Day**, the student-authors first learned about sea turtles in their native Nicaragua. Then they wrote and illustrated their own stories about sea turtles. The colorful pictures they drew reflected the colorful nature of their culture.

To celebrate the end of the writing workshop, the Felix Ruben students traveled to San Isidro de Bolas in small taxis called tuktuks. At the Turtle Party, the students wore turtle hats and played fun games as a group.

Turn the page to learn more about the Fabretto Children's Foundation and what you can do to help promote education for students like our authors in Nicaragua.

ESCOLAR PUBLICO FELIX RUBEN
Efemérides del mes de Abril

FABRETO

About Fabretto Children's Foundation

The work of Fabretto can be traced back to the actions of a man who wanted to help educate children in one of the poorest countries in Latin America. In 1948, Father Rafael Maria Fabretto traveled from Italy to Nicaragua on a mission. Seeing Nicaraguan children living in poverty, Father Fabretto committed his life to changing their lives by providing them with a home and hope for a better future through education.

Today, the Fabretto organization works with approximately 10,000 children and their families in six centers and more than fifty public elementary schools. All of this because one person decided to take action and try to make a difference.

The mission of Fabretto is a powerful one: to serve poor children over an extended period of time to address the root causes of poverty. To break the cycle of poverty, Fabretto has a number of programs to promote education, nutrition, and community development. For some students, the lunchtime meal they get at their Fabretto school is their only meal each day. By design, Fabretto creates safe, nurturing school environments so that students are healthier and have the opportunity to better their lives.

To learn more about the Fabretto Children's Foundation's work and impact, please visit www.fabretto.org. Continue reading to meet Patricia Aragonés, who switched careers to dedicate herself to helping the children of Nicaragua.

FABRETTO
children's foundation

A Heroine and Leader

Meet Patricia Aragonés, President of the Fundación Fabretto (Europe)

Organizations need strong leaders, and in the case of Fabretto, Patricia Aragonés is exactly that. Patricia was inspired to join Fabretto after a trip to Nicaragua.

Realizing that she could help and noting the tremendous impact of the Fabretto organization, Patricia decided to leave her career as a lawyer. Eventually, she became the leader of Fabretto's programs in Europe as well as a member of the advisory board in the United States.

Patricia believes that education is extremely important. In her opinion, every child's personal development affects community development, so it is crucial to educate all children.

Patricia finds fulfillment in her work for Fabretto because the organization is focused on promoting wide-ranging, sustainable programs for entire communities. She appreciates seeing the positive impact of an organization that uses education as a tool for lifelong success. Programs that teach people how to catch fish are more valuable than programs that simply give people fish to eat.

Because of her work with Fabretto, Patricia has been able to build relationships with people from diverse backgrounds and cultures around the world. She feels privileged to have these opportunities and believes her life has been enriched because of them.

To get involved and to make a difference, please visit www.dot-to-dot-books.org.

We Need Your Help

In 2007, Kevin McCaffrey read an article about violence against women in Eastern DR Congo. Outraged and tired of inaction, Kevin decided to travel to Eastern Congo to meet survivors of the region's violence. During a performance of skits by some women survivors, Kevin was pleasantly surprised to find that they were gifted storytellers. Their creativity sparked an idea that would become Dot-to-Dot Children's Books.

Transformed by his experience, Kevin asked a few friends and one of his former teachers to help him found Dot-to-Dot. Although it was initially just a small group of friends, Dot-to-Dot did not remain small for long. Nearly 225 young women and children from eight countries joined the Endangered Species, Empowered Communities (ESEC) project by writing stories about endangered animals in their countries. Their writing was the inspiration for the dozens of volunteers who helped transform their stories into the children's books that are at the core of the ESEC project.

Dot-to-Dot Children's Books is a nonprofit organization that helps people around the world use their creativity to strengthen their communities while raising awareness about important global issues. Dot-to-Dot is also an example of the power of people joining together to support a common cause. "You want students to become citizens who work to solve real-world problems." says co-founder Paul Kim. "To work with them to publish books that make a positive impact is an amazing experience for me as a teacher."

Yet the ESEC project does not end with the publication of these books. The books themselves are simply a place to start. The true value of the ESEC project is only unlocked when readers are inspired to action by what they have read and learned. This is where we need your help. We hope you have been inspired to become a part of the ESEC story by taking action. Only through your actions can the ESEC project fulfill its potential.

What You Can Do

Spread the Word

Recommend the ESEC books to your friends and family, to local libraries, and to elementary school teachers. Each book sale will generate funding to help empower girls and protect endangered species while also raising awareness about the important issues featured in each book. Visit www.dot-to-dot-books.org to see details about how the funds will be used.

Share Your Creativity

Use your creativity to write stories and develop ideas that further the message of the ESEC project and share them with others on our website. Teachers can submit classroom materials, parents can recommend activities they have done at home, and kids can upload stories and illustrations they have created to share with others. All of these creative contributions will be available to you for free on our website along with some of our own suggestions about what you can do to help change the world.

Take Action

We hope that everyone who reads this book will commit to doing something - big or small - to help protect endangered species and empower communities in the world. Go to our website to find ideas about what you can do. Many of these ideas are simple and free.

For Educators

Connecting Students to the World: Engaging K-3 Lessons for Literacy, Science, and Social Studies from Facing the Future is a teacher lesson plan guide for grades K-3 designed to complement the ESEC books. Engage students with interdisciplinary lessons aligned with language arts, science, and social studies standards. As they explore real-world environmental and social issues, students discover how they can be involved in solutions. Download Connecting Students to the World at ww.facingthefuture.org.

Featured Organizations

Dot-to-Dot Children's Books is a not-for-profit organization that develops and publishes stories to further global education and to support community development around the world. As part of our mission, we want to raise awareness about the meaningful work of the organizations that are featured in this book.

Please join us in our proud support of their work.

Facing the Future (FTF) believes in the transformative power of education to improve lives and communities, both locally and globally. FTF engages students by making academics relevant to their lives and by empowering them to think critically, develop a global perspective, and participate in positive solutions for a sustainable future. FTF's standards-based hands-on curriculum resources are in use in all 50 U.S. states and 120 countries. Learn more at www.facingthefuture.org.

The Fabretto Children's Foundation serves impoverished Nicaraguan children and their families in underserved communities. Using strategies that support the whole child, Fabretto programming complements and enhances the existing support networks in local communities. As a result, children are able to break the cycle of poverty in their families and reach their full potential. Learn more at www.fabretto.org.

The Wildlife Conservation Society (WCS) saves wildlife and wild places worldwide. WCS does this through science, global conservation, education and the management of the world's largest system of urban wildlife parks, led by the flagship Bronx Zoo. In addition, WCS runs the New York Aquarium, Central Park Zoo, Prospect Park Zoo, and Queens Zoo. Together these activities change attitudes toward nature and help people imagine wildlife and humans living in harmony. WCS is committed to this mission because it is essential to the integrity of life on Earth. Teachers can find resources at www.bronxzoo.com/educators.aspx. To learn more about the Wildlife Conservation Society, visit www.wcs.org.